Interactive
Mathematics Program®

INTEGRATED HIGH SCHOOL MATHEMATICS

The Pollster's Dilemma

FIRST EDITION AUTHORS:
Dan Fendel, Diane Resek, Lynne Alper, and Sherry Fraser

CONTRIBUTORS TO THE SECOND EDITION:
Sherry Fraser, IMP for the 21st Century
Jean Klanica, IMP for the 21st Century
Brian Lawler, California State University San Marcos
Eric Robinson, Ithaca College, NY
Lew Romagnano, Metropolitan State College of Denver, CO
Rick Marks, Sonoma State University, CA
Dan Brutlag, Meaningful Mathematics
Alan Olds, Colorado Writing Project
Mike Bryant, Santa Maria High School, CA
Jeri P. Philbrick, Oxnard High School, CA
Lori Green, Lincoln High School, CA
Matt Bremer, Berkeley High School, CA
Margaret DeArmond, Kern High School District, CA

Key Curriculum Press

Second Edition   I M P

This material is based upon work supported by the National Science Foundation under award numbers ESI-9255262, ESI-0137805, and ESI-0627821. Any opinions, findings, and conclusions or recommendations expressed in this publication are those of the authors and do not necessarily reflect the views of the National Science Foundation.

Key Curriculum Press
1150 65th Street
Emeryville, California 94608
email: editorial@keypress.com
www.keypress.com
10 9 8 7 6 5 4 3 2 1   15 14 13 12 11
ISBN 978-1-60440-146-2
Printed in the United States
of America

Project Editors
Mali Apple, Josephine Noah

Project Administrator
Emily Reed

Professional Reviewers
Rick Marks, Sonoma State University, CA
D. Michael Bryant, Santa Maria High School, CA, retired

Accuracy Checker
Carrie Gongaware

First Edition Teacher Reviewers
Kathy Anderson, Aptos High School, CA
Dan H. Brutlag, Tamalpais High School, CA
Robert E. Callis, Hueneme High School, CA
Susan Schreibman Ford, Delhi High School, CA
Mary L. Hogan, Arlington High School, MA
Jane M. Kostik, Patrick Henry High School, MN
Brian Lawler, California State University San Marcos, CA
Brent McClain, Vernonia School District, OR
Michelle Novotny, Eaglecrest High School, CO
Barbara Schallau, East Side Union High School District, CA
James Short, Oxnard Union High School District, CA
Kathleen H. Spivack, Wilbur Cross High School, CT
Linda Steiner, Orange Glen High School, CA
Marsha Vihon, Corliss High School, IL
Edward F. Wolff, Arcadia University, PA

First Edition Multicultural Reviewers
Genevieve Lau, Ph.D., Skyline College, CA
Luís Ortiz-Franco, Ph.D., Chapman University, CA
Marilyn Strutchens, Ph.D., Auburn University, AL

Copyeditor
Brandy Vickers

Interior Designer
Marilyn Perry

Production Editor
Andrew Jones

Production Director
Christine Osborne

Editorial Production Supervisor
Kristin Ferraioli

Compositors
Kristin Ferraioli, Maya Melenchuk

Art Editor/Photo Researcher
Maya Melenchuk

Technical Artists
Lineworks, Inc., Maya Melenchuk, Kristin Ferraioli

Illustrator
Juan Alvarez, Alan Dubinsky, Tom Fowler, Nikki Middendorf, Briana Miller, Evangelia Philippidis, Paul Rodgers, Sara Swan, Martha Weston, April Goodman Willy, Amy Young

Cover Designer
Jenny Herce

Printer
Lightning Source, Inc.

Executive Editor
Josephine Noah

Publisher
Steven Rasmussen

CONTENTS

The Pollster's Dilemma—The Binomial Distribution and the Central Limit Theorem

The Pollster's Dilemma

The Binomial Distribution and the Central Limit Theorem

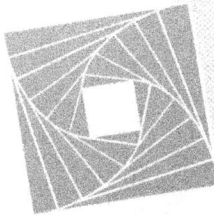

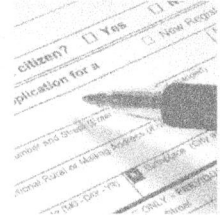

The Pollster's Dilemma—The Binomial Distribution and the Central Limit Theorem

What's a Pollster to Think?

It's one thing to take a poll of potential voters. It's quite another to determine what the results of the poll really mean.

After looking at the central problem of this unit, you'll experiment to see how much variation can occur from one poll to another, even when the polls are taken from the same population.

Vanessa Fehl lists the principles underlying good polling practices.

The Pollster's Dilemma

Many people in River City are tired of corruption in their city government. Contracts are often awarded to relatives of people in power, and politicians routinely accept bribes for favors.

Coretta Collins, who is running for mayor of River City, has pledged to end the corruption.

Everyone working on Coretta's campaign is an unpaid volunteer. Volunteers will learn about running a campaign while on the job. You are one of these volunteers, and you have been put in charge of handling Coretta's polls.

Before you began, a poll was taken of 500 eligible voters. It found that 53% of those polled intend to vote for Coretta. Of course, Coretta wants to know whether she will win the election. You need to figure out what this poll means in terms of her chances of having a majority of the people vote for her.

There are 400,000 eligible voters in River City.

1. a. How many of the people polled indicated that they would vote for Coretta?

    b. If different people had been selected for the poll, the results might have been different. How many "yes" votes (votes for Coretta) would have to be replaced by "no" votes (votes for Coretta's opponent) for the result to be a tie?

2. Based on this poll, how secure shoul Coretta feel about her lead? That is, significance can you attach to this result?

3. What information would you like to know about the poll so that you could trust it? Make a list of things you would insist on for future polls in order to have confidence in their results.

No Bias Allowed!

In everyday language, the word *bias* is often used to mean "an unfair judgment, especially against a particular racial or ethnic group." But in statistics, the word does not carry this negative connotation. Rather, **bias** refers to any built-in imbalance in a **sampling** process.

A poll with a biased sample might not give correct information about the larger population. It may slant the results in a certain direction, even if the pollster doesn't have that intention. In this activity, you will look at ways in which bias might enter into the polling process for a particular situation.

Suppose a prom committee wants to know whether $75 per ticket is too much to charge. The committee decides to poll some students and use the results to help determine whether that price is too high.

1. a. Identify the overall population in this situation.

    b. What does the term *sample* refer to in this situation?

2. Explain what might be wrong with each of these methods of choosing a sample and how each method might bias the results.

    a. Picking every tenth student who drives into the school parking lot

    b. Stopping a group of students coming out of school together and asking everyone in that group

    c. Picking one mathematics class at **random** and polling all the students in that class

3. Describe a specific plan for choosing a random sample of students in your high school. Then discuss the strengths and weaknesses of your plan.

The King's Switches

Our favorite king is tired of having his gold stolen, and he's about to leave on a vacation. He has installed an alarm system to prevent intruders from getting into his royal vault while he is gone.

The alarm system is activated and deactivated by the use of five switches. The only way to deactivate the system is to turn off all five switches.

You might think a thief could simply turn off the switches, but this isn't so easy. The system is set up so that the switches can be changed only according to these rules:

• The switch on the left may be changed (turned on or off) no matter how the other switches are set.

• Any other switch may be changed only if the switch to its immediate left is on and all other switches to its left are off.

For example, suppose the switches are currently set as shown in the diagram. You can change either switch 1 or switch 4, but you can't change switch 2, switch 3, or switch 5.

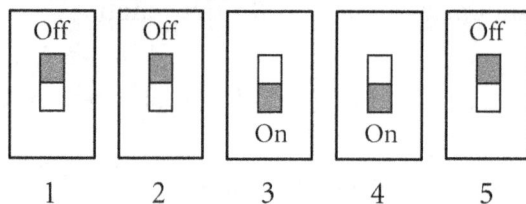

1. When the king leaves for vacation, all five switches are turned on. What is the minimum number of moves required to deactivate the alarm system?

Of course, this POW isn't only about the case of five switches, and it isn't only about getting a numeric answer.

continued

Suppose you have n switches, all turned on, and you want to turn them all off. Assume you must change switches according to the rules in the king's alarm system.

2. Find the minimum number of moves required for the cases $n = 1, 2, 3,$ and 4, and prove your results. Also prove the result you found in Question 1 for the case $n = 5$.

3. Describe a pattern you see in your results or a procedure for finding new results based on results for smaller values of n. You might also make a guess about a general formula. You can get more data by considering cases where n is more than 5 and then use this information to make guesses or test ideas.

4. Based on your pattern, procedure, or formula, find the number of moves required for $n = 10$.

5. Try to prove that your pattern, procedure, or general formula works in all cases. (That's hard!)

Write-up

In answering Questions 1 to 5, address these issues:

- The minimum number of moves required to deactivate the system for the specific cases you studied, including all cases up to $n = 5$

- Any ideas you have about a general pattern, procedure, or formula for describing the number of moves required, and an explanation of how to get the number of moves for $n = 10$

- Proofs, for each case up to $n = 5$, that your answer represents the minimum number of moves by which the system can be deactivated

Adapted from *Mathematics: Problem Solving Through Recreational Mathematics*, by B. Averbach and O. Chein. Used with permission.

Sampling Seniors

You will now examine how well polls of different sizes reflect the reality of a population.

The Scenario

The senior class at Bayside High is very lucky. They have two beautiful places where they can afford to hold their senior prom: Hamilton Hall and Cesar Chavez Center. The prom committee wants to poll the senior class to see which location members of the class prefer.

There are 150 seniors at Bayside High. Because Bayside seniors have such different schedules, it is both difficult and time-consuming to poll them all. So, the committee must decide how many people they need to poll. They want to poll as few seniors as possible but still get a good idea of what the seniors truly want.

Your Task

You will be simulating a Bayside High poll. (Actually, you will simulate many polls.) To set up the simulation, use 150 objects to represent the 150 students. The objects should be identical except for color.

continued ▶

Although the prom committee doesn't know this, 60% of the students favor Cesar Chavez Center and the remaining 40% favor Hamilton Hall. Thus, use 90 objects of one color to represent the students who favor Cesar Chavez Center and 60 objects of another color to represent the students who favor Hamilton Hall. Then put the 150 objects in a bag and mix them up.

1. a. Choose a sample size. Then, without looking into the bag, pick that many objects from your bag and record the number of votes for Cesar Chavez Center. When you are done, return all the objects to the bag and mix them up.

    b. Repeat the sampling process again and again, using the same sample size, until you have done a total of 20 polls.

2. a. Make a frequency bar graph of the results of your 20 polls.

    b. Determine how many of your polls show a majority favoring Cesar Chavez Center.

3. Repeat the entire process using a different sample size and doing another 20 polls. As time allows, continue with other sample sizes, doing 20 polls for each size.

4. Based on your results, how small a group can the committee poll and still get a good idea of what the senior class prefers? Explain your answer.

Pennant Fever Reflection

This activity looks back at ideas from the Year 3 unit *Pennant Fever* that will also play an important part in this unit.

In *Pennant Fever,* the Good Guys baseball team had seven games remaining in their regular season. For each game they played, they had a probability of .62 of winning.

1. How is the polling process in *Sampling Seniors* like the sequence of games the Good Guys played? How is it different?

2. Find the probability of the Good Guys winning all seven of their remaining games. Explain your reasoning.

3. Find the probability of the Good Guys winning exactly six of their seven remaining games. (*Reminder:* There is more than one way to win six games and lose one.)

4. Find the probability of the Good Guys winning four and losing three of their seven remaining games.

Bags of Marbles and Bowls of Ice Cream

Part I: With and Without Replacement

In *Sampling Seniors*, you completed each poll before returning the objects to the bag. This approach is called *sampling without replacement.*

Another type of sampling involves picking objects from a bag one at a time and returning each object to the bag (and mixing the objects) before picking the next object in the sample. This approach is called *sampling with replacement.*

You will now investigate each approach using two population sizes.

1. Imagine a bag of 12 marbles of which 10 are red and 2 are blue.

   a. Suppose marbles are pulled out of the bag one at a time and not put back in. (This is sampling without replacement.) If the first eight marbles pulled from the bag are red, what is the probability that the ninthmarble will be red?

   b. Imagine that the same 12 marbles are back in the bag. Again, marbles are pulled out of the bag, but this time, after each marble is selected, it is returned to the bag and mixed in with the others. (This is sampling with replacement.) If the first eight marbles pulled from the bag are red, what is the probability that the ninth marble will be red?

   c. Compare your results from Questions 1a and 1b.

2. Now imagine a bag of 12,000 marbles, of which 10,000 are red and 2000 are blue.

   a. As in Question 1a, suppose you use sampling without replacement, keeping marbles out after they are selected. If the first eight marbles pulled from the bag are red, what is the probability that the ninth marble will be red?

continued

b. As in Question 1b, suppose you use sampling with replacement, returning each marble to the bag after it is selected. If the first eight marbles pulled from the bag are red, what is the probability that the ninth marble will be red?

c. Compare your results from Questions 2a and 2b.

3. How does population size affect the distinction between sampling without replacement and sampling with replacement?

Part II: Cones and Bowls

Part II of this activity continues the review of ideas from *Pennant Fever.* You may recall the ice cream adventures of Jonathan and his sister Johanna. Jonathan likes to eat his ice cream out of a bowl, so it doesn't matter to him what order the flavors are in. For Jonathan, a bowl with a scoop of strawberry, a scoop of pistachio, and a scoop of butter pecan is the same as a bowl with a scoop of pistachio, a scoop of butter pecan, and a scoop of strawberry.

Johanna, on the other hand, eats her ice cream in cones, first eating the flavor that's on top, then the next flavor, and so on. It makes a difference to her what order the flavors are in.

4. If the ice cream store has 20 flavors and Johanna is choosing a three-scoop cone (with three different flavors), how many different cones must she consider? Explain your answer.

5. At the same store, Jonathan is choosing a three-scoop bowl (again, with three different flavors). How many different bowls must he consider? Explain your answer.

6. Explain the relationship between your answers to Questions 4 and 5.

7. Repeat Questions 4 to 6 using four-scoop cones and bowls.

Polls and *Pennant Fever*

Now that you've done some experiments with polls, it's time to look at the theory. When you take a poll of a specific size from a specific population, what's the theoretical probability of getting each possible specific result?

Based on the ideas in Part I of *Bags of Marbles and Bowls of Ice Cream,* you will be using the simplifying assumption that polling can be viewed as sampling with replacement. That assumption will allow you to apply ideas from the Year 3 unit *Pennant Fever* to find probabilities.

Maceo Sovin-Martinez applies ideas from the Year 3 unit "Pennant Fever" to help determine theoretical probabilities in this unit.

The Theory of Three-Person Polls

Imagine that you are conducting polls on behalf of a certain candidate in a two-candidate race. You will ask people whether they are going to vote for your candidate. Assume that participants give honest yes-or-no answers and that there are no undecided voters.

The outcome—the number of "yes" answers—will depend on how opinions are divided in the overall population and, of course, on the size of your poll. For polls of a given size, each possible outcome has a specific theoretical probability of occurring.

In this activity, you will consider the case of three-person polls from a specific overall population and find the theoretical **probability distribution** for the possible outcomes. That is, imagine that you are taking many separate polls and that you question three people in each poll.

continued ▶

Begin by making this assumption:

- The overall population is 60% in favor of your candidate.

This figure of 60% (or .6) is referred to as the **true proportion.**

For simplicity, make these two additional assumptions:

- Each voter polled is picked at random from the total voting population.
- The overall population is big enough compared to the sample size that you can treat the problem as if it involves sampling with replacement.

These three assumptions can be combined into one:

> *Every voter picked has a probability of .6 of being in favor of your candidate.*

Answer the questions based on this combined assumption.

1. List the possible outcomes for the number of "yes" votes in such a poll.

2. Find the probability of each possible outcome.

3. Make a probability bar graph showing your results.

Graphs of the Theory

In *The Theory of Three-Person Polls,* you examined what happens if a three-person poll is taken from a population that is 60% in favor of a given candidate. In Questions 1 and 2 of this activity, you'll consider what happens as the true proportion changes. Question 3 asks you to think about the effect of changing the sample size. Continue to use a "sampling with replacement" model to find the probabilities.

1. Suppose the true proportion, which we usually call p, is .55. That is, suppose 55% of the overall population is in favor of a given candidate. The fraction of votes in favor of the candidate in a specific poll is called the **sample proportion.** We often use the symbol $\hat{p}$ for the sample proportion. (This symbol is read as "p hat.") Make a probability bar graph showing the probability of getting each possible value of $\hat{p}$ for a three-person poll.

2. Now make a probability bar graph for the case in which the true proportion is 70%. Compare this graph with the graph from Question 1.

3. Generalize your work from Questions 1 and 2. That is, assume the true proportion is p, and find the probability for each possible sample proportion for a three-person poll.

   Your probabilities will be expressions in terms of p. You do not need to make a probability bar graph for this general case.

4. In Question 1, you made a probability bar graph for a three-person poll using a true proportion of 55%. How do you think the graph would change if the sample size were increased (keeping the same true proportion of 55% for the overall population)?

The Theory of Polls

In *The Theory of Three-Person Polls,* you found the theoretical probability for each possible outcome of a three-person poll. As the person in charge of polls for Coretta Collins, you need to study the theory of polling more fully in order to better understand the reliability of polls.

For larger polls, there are more possible outcomes, each with a theoretical probability. The main focus of this activity is on how the probabilities change as the poll size changes.

For the sake of making comparisons, assume throughout this activity that the true proportion is .6. That is, assume 60% of the population favors the candidate.

continued ▶

Once a poll is taken, the pollster can compute the sample proportion, which is the fraction of those polled who favor the candidate. For the case of a three-person poll, the theoretical distribution for the sample proportion is shown in the graph. Notice that roughly 35% of such polls (.064 + .288) would show the candidate is trailing, even though the candidate has the support of 60% of the overall population.

Distribution of 3-person polls with true proportion = 60%

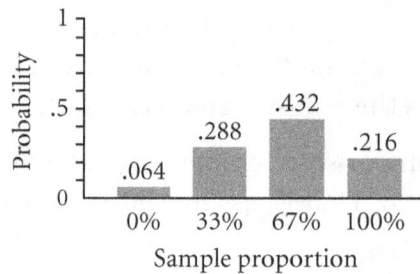

1. Consider the case of a five-person poll. The number of voters in the poll who support the candidate could be 0, 1, 2, 3, 4, or 5, so the sample proportion could be 0%, 20%, 40%, 60%, 80%, or 100%.

   a. Find the probability of each of these possible results. (*Reminder:* Assume the true proportion is .60.)

   b. Make a probability bar graph of your results.

   c. What percentage of five-person polls correctly show the candidate leading?

2. Now consider the case of a nine-person poll.

   a. Find the probability of each possible outcome.

   b. Make a probability bar graph of your results.

   c. What percentage of nine-person polls correctly show the candidate leading?

Civics in Action

The senior civics class is about to put theory into practice by conducting the election of the senior class president. Clarence is one of two candidates running for this office.

There are 60 members of the senior class. Although nobody knows it yet (except you), 50 of these 60 students intend to vote for Clarence.

The editor of the class newspaper, Clarissa, plans to take an exit poll so she can publish a prediction before the votes are officially counted. An exit poll is a poll of people as they leave the voting booth. Clarissa intends to predict only the winner, not how many votes the winner will get.

Clarissa needs to study for an important physics test. Therefore, although she will pick students randomly from a list and get their votes in an anonymous way, she probably won't actually ask many students how they voted.

1. Suppose Clarissa asks only one person and uses that person's vote to make her prediction. What is the probability that she will get a result that correctly predicts the winner?

2. Suppose Clarissa polls three people.
   a. Make a probability bar graph for the possible outcomes of this three-person poll.
   b. What is the probability that her poll will correctly predict the winner? In other words, what is the probability that the poll will show a majority supporting Clarence?

3. Pick at least one more poll size (greater than 3) and make a probability bar graph for the possible outcomes. Also find the probability that a poll of that size would correctly predict the winner.

4. If it were up to you, what is the least number of people you would be satisfied to poll (assuming you didn't know the actual number of students in favor of Clarence)? Give some reasons for your answer.

Normal Distributions Revisited

Using the "sampling with replacement" model, you've seen two powerful facts about polling:

- Poll results follow a binomial distribution.
- For large polls, binomial distributions look a lot like normal distributions.

Because of these two facts, the normal distribution will play a key role throughout the rest of this unit. In the upcoming activities, you'll review and extend what you learned in previous years about the normal distribution.

Yaravi Anaya prepares the graphs she'll present to the class comparing normal curves with different means and standard deviations.

The Central Limit Theorem

The **central limit theorem** is a powerful and general principle about statistics and probability. Although it applies to many situations, only one aspect of the theorem will be used in this unit. In a very succinct form, this special case says:

As n gets bigger and bigger, the probability distribution for an n-person poll looks more and more like the normal distribution.

Explaining more fully what this principle means requires some discussion of the distribution for n-person polls and the normal distribution.

The *n*-person Poll and the Binomial Distribution

Start with this principle:

If n people are sampled from a population, and the size of the total population is much larger than n, we can treat the poll as if it were a case involving sampling with replacement.

According to this principle, polling people on how they will vote in a two-candidate election is similar to repeatedly flipping a coin (a not-necessarily balanced coin) or to playing a sequence of baseball games in which a team has a fixed chance of winning each game. For each person polled, there are two possible outcomes, which we might call "yes" and "no." The probabilities of these two outcomes are the same for every person polled.

Suppose p represents the fraction of "yes" voters in the overall population (usually called the *true proportion*). Then, for each person polled, the probability of a "yes" vote is p and the probability of a "no" vote is $1 - p$.

The number of "yes" votes in an n-person poll can be any number from zero through n. The probability distribution for the poll tells what the chances are of getting each result. Specifically:

The probability of getting exactly r "yes" votes in an n-person poll is $_nC_r \cdot p^r \cdot (1 - p)^{n-r}$.

continued ▶

This collection of probabilities, for the different values of r, is called the **binomial distribution.** This is a *discrete* distribution, and the set of possible outcomes is finite.

The binomial distribution is different for each choice of n and p. In the general sense, we often refer to n as the number of "trials" and p as the probability of "success."

In comparing the binomial distribution to the normal distribution, we consider the *fraction* of "yes" votes in the sample, rather than the number of such votes. This fraction, usually called the *sample proportion,* is simply $\frac{r}{n}$. It is often expressed as a percentage, and it can be anywhere from 0% to 100%.

For any particular binomial distribution, the set of possible values within this range is finite. For instance, if $n = 4$, the sample proportion must be one of these values: 0%, 25%, 50%, 75%, or 100%.

The Normal Distribution

The normal distribution is an example of a *continuous* distribution, because any numeric outcome is possible. Continuous distributions are described by giving the probability of getting an outcome (often called a *result*) within any interval, rather than by giving the probability of getting each particular result.

The first diagram below shows an example of a **normal curve.** (The equation for this curve is discussed in the activity *Graphing Distributions.*) The horizontal scale measures possible numeric results, such as the proportion of votes for a particular candidate. The vertical scale corresponds to the frequency of the various outcomes. The probability of getting a result between a and b is equal to the area of the shaded region. The total area under the curve is assumed to equal 1.

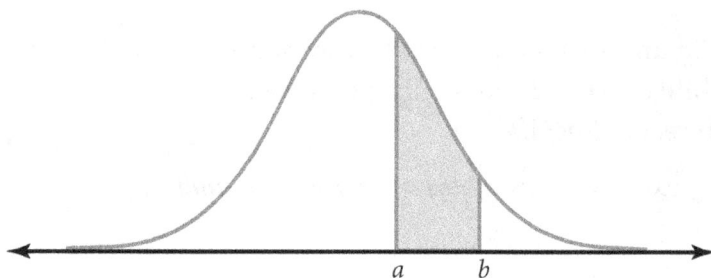

continued ▶

As with the binomial distribution, there are many different normal distributions. And like the binomial distribution, the normal distribution depends on two parameters. For the normal distribution, these two parameters are the *mean* and the *standard deviation.*

The *mean* gives the average result for the distribution, and the normal curve is symmetric about this value. The mean is often represented by the Greek letter μ ("mu"), and it can be positive, negative, or zero. The *standard deviation* tells how "spread out" the curve is. The standard deviation is generally represented by the Greek letter σ ("sigma"), and it can be any positive number.

This diagram shows two normal curves with the same standard deviation but different means:

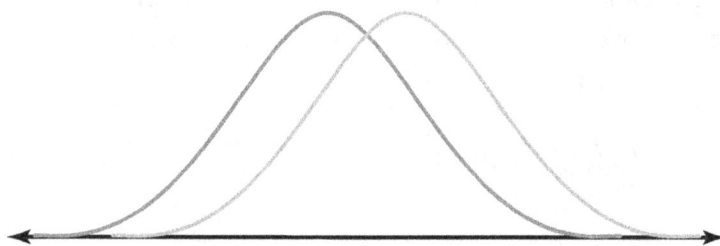

This diagram shows two normal curves with the same mean but different standard deviations:

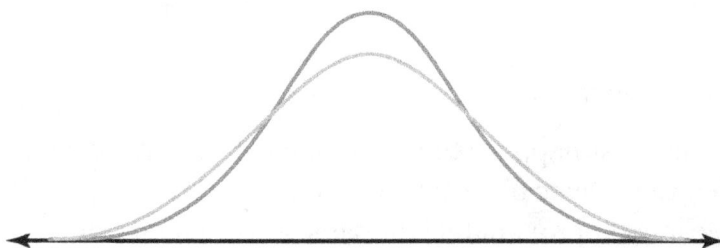

One useful aspect of the normal distribution is that the probability of getting a result within a given interval depends only on how the endpoints of the interval relate to the mean and standard deviation. For example, in the next diagram, the shaded area extends from one standard deviation below the mean to one standard deviation above the mean—that is, from $\mu - \sigma$ to $\mu + \sigma$. No matter what the values of μ and σ, this area is about 68% of the total area under the curve.

continued ▶

Thus, if an experiment follows a normal distribution with mean μ and standard deviation σ, then in the long run, approximately 68% of all results from the experiment will fall between $\mu - \sigma$ and $\mu + \sigma$.

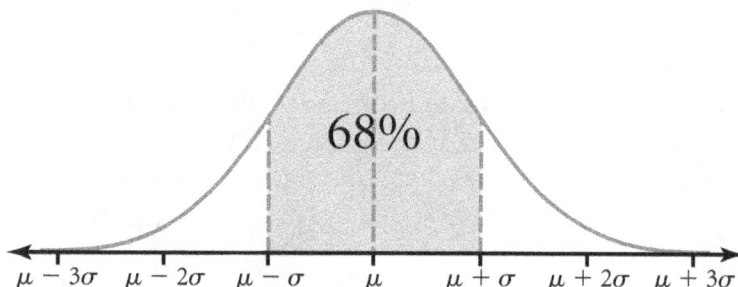

In the next diagram, the shaded area includes all results within two standard deviations of the mean. This area is about 95% of the total. In other words, in the long run, approximately 95% of all results lie between $\mu - 2\sigma$ and $\mu + 2\sigma$. Again, this is true for every normal distribution.

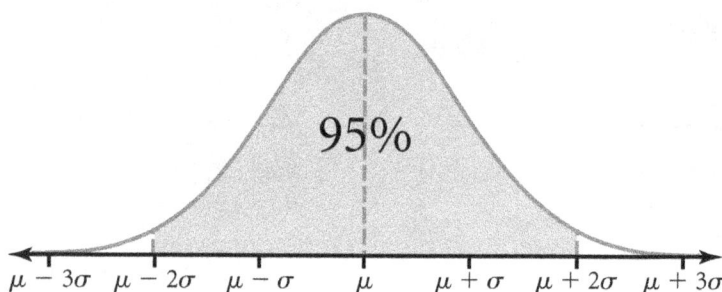

The Central Limit Theorem

The central limit theorem tells what happens to the binomial distribution as n gets larger. It states that the chance of getting a sample proportion within a certain interval becomes approximately the same as the chance of getting a result within the same interval for a certain normal distribution.

The value of μ (the mean) and the value of σ (the standard deviation) for the approximating normal distribution depend on n (the poll size) and p (the true proportion). Later, you will find formulas for μ and σ in terms of n and p.

For a fixed value of p, as n gets bigger and bigger, the normal distribution becomes a better and better approximation for the binomial distribution. In fact, no matter how good an approximation you want, you can find a value of n big enough to give an approximation that is that good.

Deviations of Swinging

A group of students is trying to find out how long it took for 12 complete swings of a 30-foot pendulum. They repeatedly measure the time for 12 swings and get a variety of results. They determine that the mean of their results is 72 seconds and that the standard deviation is 1.5 seconds.

Assume that measurements of the pendulum's period fit a normal distribution.

1. What percentage of future measurements should give results between 72 and 75 seconds?

2. What percentage of future measurements should give results greater than 73.5 seconds?

3. What percentage of future measurements should give results greater than 69 seconds?

Means and More in Middletown

For each of Questions 1 to 3, assume the situation involves a normal distribution. Question 4 asks you to comment on these assumptions.

In each case, sketch a normal distribution, identifying and labeling the positions that are one and two standard deviations from the mean. Use your sketches to help justify your answers.

1. An analysis of apartments in Middletown shows a mean rent of $650 per month, with a standard deviation of $150 per month. What are three conclusions you can draw from this information?

2. The Middletown Police Department gives a standardized test to potential captains. The mean score is 270, with a standard deviation of 15 points. A score of 300 points is required to pass. Approximately what percentage of those who take the test will pass?

3. The Middletown Track Club surveyed some of its members to get an idea of how they train. The survey shows that members run an average of 35 miles per week, with a standard deviation of 7 miles per week. Approximately what percentage of the runners in the club run less than 42 miles per week?

4. For each of Questions 1 to 3, discuss whether it is reasonable to assume the situation involves a normal distribution.

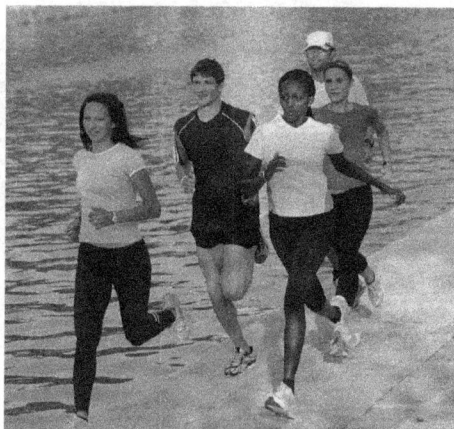

Graphing Distributions

The normal distribution is a very special set of probabilities for the possible outcomes of an experiment. In a normal distribution, any numeric result is possible, although not all results are equally likely. The probability of getting a result within a given interval is determined by a complex formula. Specifically, for a normal distribution with mean μ and standard deviation σ, the normal curve is the graph of this equation, in which y is a function of x:

$$y = \left(\frac{1}{\sigma\sqrt{2\pi}}\right) \cdot e^{-\frac{1}{2}\left(\frac{x-\mu}{\sigma}\right)^2}$$

The probability that a result is between a and b is the area under the normal curve between the vertical lines $x = a$ and $x = b$.

The simplest normal curve is the case $\mu = 0$ and $\sigma = 1$, for which the equation is

$$y = \left(\frac{1}{\sqrt{2\pi}}\right) \cdot e^{-\frac{1}{2}x^2}$$

This special case is sometimes called the *standard* (or *normalized*) normal curve. Principles about normal distributions are often expressed in terms of this special case.

In the general case, we have the initial factor $\frac{1}{\sigma\sqrt{2\pi}}$, so that the total area under the normal curve is equal to 1, no matter what μ and σ are. Having the total area equal to 1 corresponds to having the sum of all the probabilities equal to 1.

1. Use a calculator to show the standard normal curve, which is the case in which $\mu = 0$ and $\sigma = 1$.

2. Use a calculator to show the normal curve for $\mu = 2$ and $\sigma = 3$.

continued ◗

3. Graph normal curves with $\mu = 2$ but with standard deviations other than 3. What changes do you see?

4. Graph normal curves with $\sigma = 3$ but with means other than 2. What changes do you see?

Gifts Aren't Always Free

Craig's aunt is buying him a car for his high school graduation. He lives in a rural area, so a car will be very handy.

Craig needs to work this summer, but he'd also like to have some free time. He wants to figure out how many hours he needs to work. He knows how much money he needs for everything except gasoline—he has figured out that he will be driving about 500 miles per week, and he knows that gasoline in his area costs about $2.75 per gallon.

He has also done some research on the kind of car he is getting. He found a study that rates the fuel efficiency for each of a number of cars of this particular model and year. According to the study, the average of these individual ratings is 23 miles per gallon, with a standard deviation of 2.5 miles per gallon. Craig has not studied statistics and doesn't know what standard deviation means.

1. Explain to Craig what it means that the standard deviation is 2.5 miles per gallon.

2. Tell Craig how much he should plan on spending for gasoline per week, and why. Assume the variation in fuel efficiency among cars of this model is normally distributed.

3. Explain to Craig how normal distributions fit into all of this.

Normal Areas

The diagram shows the graph of the equation for the standard normal curve, which is

$$y = \left(\frac{1}{\sqrt{2\pi}}\right) \cdot e^{-\frac{1}{2}x^2}$$

This is the normal curve with mean 0 and standard deviation 1. The total area under this curve is equal to 1.

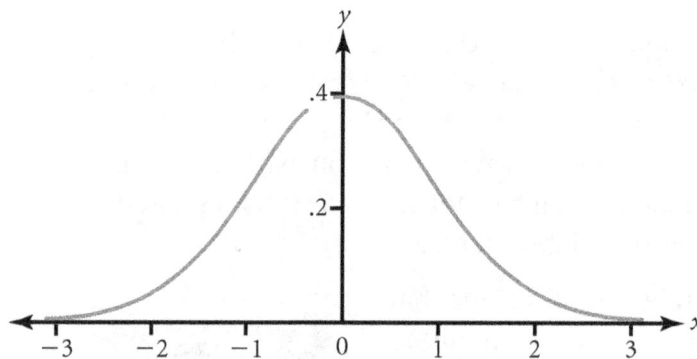

You know that the area under the curve between $x = -1$ and $x = 1$ is approximately .68, or 68% of the total area under the curve. You also know that the area between $x = -2$ and $x = 2$ is approximately .95, or 95% of the total.

Use these facts and the graph to estimate the answers to these questions, which concern other areas and percentages.

1. What percentage of the area is between $x = -0.5$ and $x = 0.5$?

2. What percentage of the area is between $x = -1.5$ and $x = 1.5$?

3. What interval symmetric about $x = 0$ will contain 50% of the area?

4. What interval symmetric about $x = 0$ will contain 80% of the area?

The Normal Table

You have seen that for all normal curves, the area between $x = \mu - \sigma$ and $x = \mu + \sigma$ is approximately .68, no matter what the values of the mean μ or the standard deviation σ are. Similarly, the areas of other regions symmetric about the mean depend only on how many standard deviations they extend from μ.

Mathematicians have computed these areas precisely. Because these computations are not easy, the results are often put into a table for easy reference.

The table on the next page gives the areas under the normal curve for various regions that are symmetric about the mean. In the table, z represents the number of standard deviations that the area extends in each direction. The most commonly used values, for $z = 1$ and $z = 2$, are shown in bold type.

For example, the case $z = 2$ refers to the area between $x = \mu - 2\sigma$ and $x = \mu + 2\sigma$ (that is, the area within two standard deviations of the mean). The table entry for $z = 2$ is .9545, which corresponds to the fact that this area is about 95% of the total area.

Similarly, the table value .3829 for $z = 0.5$ means that about 38% of the area is between $x = \mu - 0.5\sigma$ and $x = \mu + 0.5\sigma$ (that is, within half of one standard deviation of the mean).

continued ▶

This table gives the areas under the normal curve for various regions that are symmetric about the mean.

z (number of standard variations)	Area within z standard deviations of the mean	z (number of standard variations)	Area within z standard deviations of the mean
0	0	1.7	.9109
0.1	.0797	1.8	.9281
0.2	.1585	1.9	.9426
0.3	.2358	**2.0**	**.9545**
0.4	.3108	2.1	.9643
0.5	.3829	2.2	.9722
0.6	.4515	2.3	.9786
0.7	.5161	2.4	.9836
0.8	.5763	2.5	.9876
0.9	.6319	2.6	.9907
1.0	**.6827**	2.7	.9931
1.1	.7287	2.8	.9949
1.2	.7699	2.9	.9963
1.3	.8064	3.0	.9973
1.4	.8385	4.0	.99994
1.5	.8664	5.0	.9999994
1.6	.8904		

More Middletown Musings

The situations in this activity are similar to those in *Means and More in Middletown*, but this time you will need the detailed information about the normal curve from *The Normal Table*. As before, assume the data set for each situation is normally distributed.

1. You might recall that an analysis of apartments in Middletown shows a mean rent of $650 per month, with a standard deviation of $150 per month. Approximately what percentage of renters pay between $530 and $770 per month?

2. The Middletown Police Department decided to raise the test score necessary to be eligible to become captain to 306 points. The mean score is 270, with a standard deviation of 15 points. Based on the new passing score, approximately what percentage of those who take the test will pass?

3. Against the best wishes of its membership, the executive board of the Middletown Track Club decided to pick a member at random to represent the club in the state championship. Most members believe that the person representing the club should be someone who runs at least 40 miles per week. What is the probability that a member chosen at random will meet this criterion? Recall that the club's survey shows that members run an average of 35 miles per week, with a standard deviation of 7 miles per week. (*Note:* This question involves a *z*-value that is not in *The Normal Table*, so estimate the probability based on values that *are* in the table.)

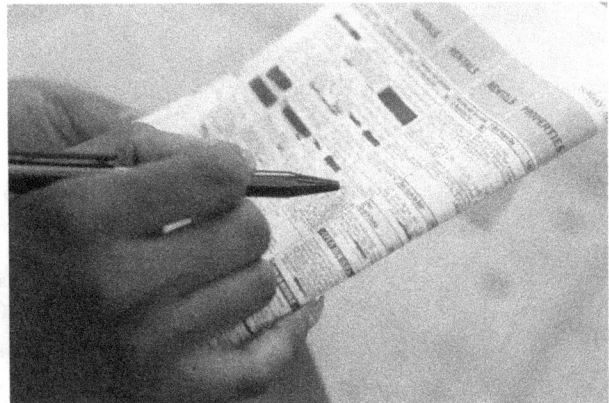

Back to the Circus

You may recall the next scenario from the Year 1 unit *The Pit and the Pendulum.*

> A circus performer wants to ride her bicycle right up to a brick wall and stop dramatically, very close to the wall and without crashing. She needs to know when to apply the brakes.

Students were asked to propose a plan to collect and analyze data that would help the performer make a prediction about when to apply the brakes. One student proposed this plan:

> The performer should do an experiment. She should draw a line on the floor. Then she should ride at full speed toward the line and apply her brakes just when she reaches the line. She should do this as a controlled experiment, so that she always rides at the same speed, applies her brakes with the same pressure, has the same amount of air in her tires, and so on. Each time she stops, she should measure how far she goes past the line.

> She should use her results to tell her how far from the wall she should stop.

Suppose the circus performer carries out this experiment many times and finds that the mean of her stopping distances is 2 meters and the standard deviation is 0.3 meter.

If the performer is willing to hit the wall 5% of the time, how far away from the wall should she apply her brakes? (Presumably, on the rare occasions when she hits the wall, she'll be going very slowly at the time of impact and won't get hurt.)

Gaps in the Table

The Normal Table tells you the probability of getting a result within a given range for a data set that is normally distributed. For instance, for $z = 0.6$, the table gives a value of .4515, which means there is approximately a 45% chance that a given result will lie within 0.6 standard deviation of the mean.

Unfortunately, the table lists only values for z that are multiples of 0.1. What do you do about in-between values, such as $z = 0.65$? That is, for normally distributed data, what is the probability of getting a result within 0.65 standard deviation of the mean?

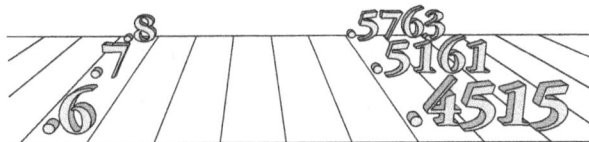

The Intuitive Idea

Intuition might lead you to reason something like this:

The value $z = 0.65$ is exactly halfway between $z = 0.6$ and $z = 0.7$. The value $z = 0.6$ has an associated probability of .4515, and the value $z = 0.7$ has an associated probability of .5161. So, we might expect the probability associated with $z = 0.65$ to be exactly halfway between .4515 and .5161.

1. Test out this intuitive idea.

   a. Find the probability associated with $z = 1.8$ and the probability associated with $z = 2.0$.

   b. Find the number exactly halfway between these two probabilities.

   c. Compare your answer to part b with the probability in the table for $z = 1.9$. Are they exactly the same? Are they close?

What Actually Happens

What you should have found is that the probability in the table for $z = 1.9$ is not exactly halfway but is pretty close to halfway. It's actually a little closer to the probability for $z = 2.0$ than to the probability for $z = 1.8$.

continued ▸

This example is a special case of a general approximation technique called *linear interpolation*. Although this method may not give the exact value, it often gives a good approximation. This technique is especially useful for functions that can't be computed in any simple way, such as the function giving probabilities associated with a given z-value.

2. Here's an illustration of this technique with a function that's much simpler than the one represented by the normal table. Consider the "squaring" function, that is, the function whose equation is $y = x^2$. This could be represented by the notation $f(x) = x^2$.

   a. Find y when $x = 7$. In other words, find $f(7)$.

   b. Find y when $x = 9$. In other words, find $f(9)$.

   Using linear interpolation, you would expect $f(8)$ to be about halfway between $f(7)$ and $f(9)$.

   c. Check this out. That is, find the number halfway between $f(7)$ and $f(9)$, and compare that number to $f(8)$.

   d. Use a graph of the function $y = x^2$ to explain why the number halfway between $f(7)$ and $f(9)$ is either higher or lower than $f(8)$.

3. Now examine what happens if you apply this technique to the function g defined by the equation $g(x) = 2x - 1$.

   a. Find $g(7)$ and $g(9)$, and then find the number halfway between these results.

   b. Compare your answer with the value of $g(8)$.

   c. What is it about the function g that makes the technique give the exact value in this case?

4. One strength of the technique of linear interpolation is that it isn't limited to "halfway" points.

   a. Use linear interpolation to estimate the probability associated with $z = 0.82$.

   b. Use linear interpolation to estimate the probability associated with $z = \frac{4}{3}$.

A Normal Poll

Clarence's younger brother, Henry, is one of two candidates running for president of his middle school. As it happens, Clarissa's younger sister, Harriet, is editor of the school newspaper.

Although neither Harriet nor Henry knows it, 60% of the student body plans to vote for Henry.

Harriet is more industrious than her big sister. She plans to take a 50-person poll at the school, selecting students at random. Based on which candidate gets a majority in her poll, Harriet will make a prediction about who will win the election.

For a poll of this size, the normal distribution gives a very good approximation for finding the probabilities of different outcomes (assuming the overall population is large enough). The normal distribution that fits this situation best has mean $\mu = .6$ and standard deviation $\sigma = .069$.

1. Does it make intuitive sense that the mean for this normal distribution is .6? Why?

2. Use the normal approximation to find the probability that Harriet's poll will show Henry as the winner.

3. One guideline says that if you want to use sampling with replacement as a model for polling, your poll size should be no more than 5% of the overall population. How big must the student population be if Harriet's poll fits this guideline?

A Plus for the Community

The Community Recreation Center is holding a fair to get people more involved in the center and to raise some money.

One of the games at the fair is a "wheel of fortune." For each ticket you buy, you spin the wheel once. A spin of the wheel will win you $1, $2, $3, or $10.

The $1 and $2 results each have a probability of .4. The $3 result has a probability of .15. The $10 result has a probability of .05.

1. Suppose the center sells 1000 tickets for the game and the results match the theoretical probability distribution perfectly. That is, suppose 400 people win $1; 400 people win $2; 150 people win $3; and 50 people win $10. What will be the average amount the center pays out per spin?

2. Would the answer to Question 1 be different if the center had sold 5000 tickets? What about other numbers of spins? Explain.

Means and Standard Deviations

The central limit theorem guarantees that under suitable assumptions about poll size, the theoretical distribution of poll results approximately follows a normal curve. But which normal curve?

There is a different normal distribution for each possible mean and standard deviation. Your next task is to relate these two parameters to the size of the poll and to the true proportion for the population from which the sample is taken.

Trey Thompson reviews the steps for computing standard deviation in preparation for extending this concept.

Mean and Standard Deviation for Probability Distributions

For any given set of data, the *mean* of the set is simply the average. That is, you add the items in the data set and divide by the number of items.

To find the *standard deviation* for a set of data, follow these steps:

a. Find the mean of the data items.

b. Find the square of the difference between each data item and the mean.

c. Add the squared differences.

d. Divide by the number of items.

e. Take the square root of this quotient.

The terms *mean* and *standard deviation* can be defined in a similar way for a probability distribution with a finite number of outcomes. The definition involves two stages:

• Creating a set of data that exactly fits the distribution

• Finding the mean and standard deviation for this data set

The mean and standard deviation for the probability distribution are defined to be the mean and standard deviation of this set of data.

It's true that there can be different data sets, of different sizes, that exactly fit the probability distribution. But you can use the distributive property to prove that all such data sets will have the same mean and the same standard deviation. So defining mean and standard deviation in terms of one particular data set does make sense.

In fact, the distributive property shows that you can use the probabilities directly, without creating a data set, to compute both the mean and the standard deviation. In many statistics books, the mean and the standard deviation are defined that way.

continued ▶

The definitions of mean and standard deviation can be expressed as shown here, using summation notation:

mean of a discrete probability distribution $= \sum P(x_i) \cdot x_i$

standard deviation of a discrete probability

$$\text{distribution} = \sqrt{\sum P(x_i) \cdot (x_i - \mu)^2}$$

The variable x_i represents the possible outcomes for the distribution. $P(x_i)$ represents the probability of outcome x_i.

As with data sets, the mean and standard deviation are often represented by the symbols μ and σ. The **variance** (for either a set of data or a probability distribution) is simply the square of the standard deviation. It is often represented as σ^2.

A Distribution Example

You've seen that the terms *mean* and *standard deviation* can be defined for a probability distribution in ways that are similar to their definitions for data sets. You will now look at an example of how that is done.

Consider the experiment of flipping three coins and counting the number of heads. There are four possible outcomes: no heads, one head, two heads, and three heads. The probability distribution for this three-coin experiment can be expressed by these equations:

- $P(0) = \dfrac{1}{8}$

- $P(1) = \dfrac{3}{8}$

- $P(2) = \dfrac{3}{8}$

- $P(3) = \dfrac{1}{8}$

The simplest set of data that fits this distribution is 0, 1, 1, 1, 2, 2, 2, 3—that is, flipping the three coins eight times and getting no heads once, getting one head three times, getting two heads three times, and getting three heads once.

1. Find the mean for the data set 0, 1, 1, 1, 2, 2, 2, 3.

2. Find the standard deviation for this data set.

3. Find the mean for the three-coin probability distribution using the formula

$$\text{mean} = \sum P(x_i) \cdot P(x_i)$$

where the variable x_i represents the possible outcomes of the experiment and $P(x_i)$ represents the probability of outcome x_i. In this context, x_i represents the outcomes 0, 1, 2, and 3, so the summation means $P(0) \cdot 0 + P(1) \cdot 1 + P(2) \cdot 2 + P(3) \cdot 3$.

continued ▶

4. Explain why the formula for the mean in Question 3 gives the same result you found in Question 1.

5. Standard deviation can be found using the formula

$$\text{standard deviation} = \sqrt{\sum P(x_i) \cdot (x_i - \mu)^2}$$

   a. Write an expression, without using summation notation, for what this formula means in the case of the three-coin probability distribution.

   b. Find the numeric value of your expression.

6. Explain why the formula in Question 5 gives the same result you found in Question 2.

The Search Is On!

You now know how the concepts of mean and standard deviation are defined for theoretical probability distributions.

In this unit, the focus is on probability distributions that describe the results of polls of different sizes. We are assuming that the overall population is large compared to the size of the poll, so we are analyzing the probabilities as if each poll is done using sampling with replacement.

Based on this assumption, the probabilities for different poll results fit the binomial distribution, which depends on two parameters:

- n, the number of trials (For polls, this is the size of the poll.)
- p, the probability of success (For polls, this is the fraction supporting the candidate in the overall population.)

In this activity, your goal is to find formulas for the mean and standard deviation of the binomial distribution in terms of these two parameters. You will first find the mean and standard deviation for specific values of p and n, and then look for patterns. Here are two suggestions to help you find patterns in the data:

- Look at the *number* of votes a candidate gets in the poll rather than the *proportion* of votes the candidate gets.
- Look for a formula for the *variance* rather than the *standard deviation*. (*Reminder:* The variance is the square of the standard deviation.)

Once you have a formula for the variance, you will use it to get a formula for the standard deviation.

Your Task

Follow these steps to develop your formulas:

1. Choose a value for p (the fraction of voters in favor of the candidate). For your chosen value of p, choose several values for n, find the mean and the variance of the number of "yes" votes for each of those values of n, and then develop formulas for the mean and variance in terms of n.

continued ▶

2. Once you have mean and variance formulas in terms of n for one value of p, get formulas for other values of p.

3. Find a general formula for the mean and variance of the number of "yes" votes in terms of n and p.

4. Use your variance formula to write a formula for standard deviation.

Why Is That Batter Sneezing?

If polling is done for a two-person election using sampling with replacement, then the poll consists of independent repetitions of an event with only two possible outcomes. As you have seen, the set of probabilities associated with this type of polling is called the *binomial distribution.*

The same basic theory applies to many other situations as well. Questions 1 and 2 each describe a situation that involves the binomial distribution.

1. A baseball player is batting .400, which means he has gotten a hit in 40% of his previous times at bat. (Ignore the complication of bases on balls and other situations in which a turn at bat doesn't figure into a player's batting average.)

   Based on the player's record, assume that for each future time at bat, the player has a probability of .4 of getting a hit.

   a. If he bats three times in the next game, what is the probability that he will get exactly one hit? Exactly two hits? Exactly three hits? No hits at all?

   b. Make a probability bar graph showing your results.

continued ▶

2. Alida has the flu, but she has an important meeting today with five other people. She is concerned about giving them the flu, because she is in the most contagious stage of the disease. Suppose that according to medical information about this type of flu, each person at Alida's meeting has a 30% chance of catching the flu from her.

   a. What are the chances that no one at the meeting will become infected?

   b. Find the probability that the number of people infected is exactly one, exactly two, and so on.

   c. Make a probability bar graph showing your results.

   d. Suppose no one else at the meeting got infected. Would you still believe that Alida was in the most contagious stage? Explain.

3. Make up another question that you find interesting and that involves the binomial distribution.

Putting Your Formulas to Work

In *The Search Is On!* you found formulas for the mean and standard deviation for the binomial distribution. If you use sampling with replacement, these formulas apply to the probability distribution of poll results.

You also know that if *n* is "big enough," the binomial distribution can be approximated by a normal distribution. Not surprisingly, this normal distribution has the same mean and standard deviation as the binomial distribution it approximates. This means you can use the formulas from *The Search Is On!* for this normal distribution.

In Question 1, you will put these formulas to use. Question 2 asks a more general question about the formula for standard deviation.

1. Remember that you are in charge of polls for Coretta. According to the poll that was taken before you started, she has approximately 53% of the vote. In other words, the sample proportion $\hat{p}$ in that poll was .53.

   You are about to take a new poll, and you want the results to be encouraging. If it looks like Coretta's ahead, people may be more willing to contribute to or volunteer to work on the campaign.

   a. Suppose Coretta really does have 53% of the vote (so the value of $\hat{p}$ from the earlier poll is exactly correct). If you now do a 300-person poll, what is the probability that your poll will show her to be in the lead? Explain how to use the normal distribution and the formulas from *The Search Is On!* to answer this question.

   b. What would be the answer to part a if you did a 600-person poll instead? Explain.

continued

2. Look at your formula for the standard deviation of the number of "yes" votes in an *n*-person poll. According to the formula, the standard deviation increases as *n* gets bigger. This means that as the poll size increases, so does the fluctuation.

On the other hand, intuition suggests that a bigger poll should be more accurate. Doesn't this mean there should be less fluctuation among big polls than among small ones? How do you explain the fact that the standard deviation is bigger for bigger polls?

From Numbers to Proportions

You know that the important thing in an election is the proportion of votes that a candidate gets, not the actual number of votes. But the formulas found in *The Search Is On!* were for the mean and standard deviation of the *number* of votes for a given candidate in a poll.

In this activity, you will find formulas for the mean and standard deviation of the *proportion* of votes for a candidate in a poll.

1. a. Imagine that you have done six 500-person polls for a two-person election. Make up results for these polls, giving the number of votes in favor of your candidate in each poll.

   b. Use your six results to find the mean and standard deviation of the number of votes in favor of your candidate.

2. a. Compute the *proportion* of votes for your candidate for each of your six results from Question 1a.

   b. Find the mean and standard deviation for this set of six proportions.

3. Compare your results for Questions 1b and 2b.

   a. How does the mean for the *number* of votes compare to the mean for the *proportion* of votes?

   b. How does the standard deviation for the *number* of votes compare to the standard deviation for the *proportion* of votes?

Now use what you have learned to tackle the main objective.

4. Find formulas for the mean and standard deviation for the *proportion* of votes the candidate gets in an n-person poll (in terms of n and p, where p is the true proportion).

Is Twice as Many Twice as Good?

Your growing expertise in polling has caught the eye of a candidate for city council. He has a limited budget for polling, and you figure that given the funding, you can do a 200-person poll.

Based on previous elections, the candidate is fairly sure he has the support of 60% of the overall population. He would like to see a poll confirm this belief.

You point out that even if his overall support really is 60%, you can't guarantee that a poll will give this exact result. In response, he asks you to give him an interval within which you think the poll result will occur.

1. If the candidate really does have the support of 60% of the population, what interval around 60% has a 95% chance of containing your poll result?

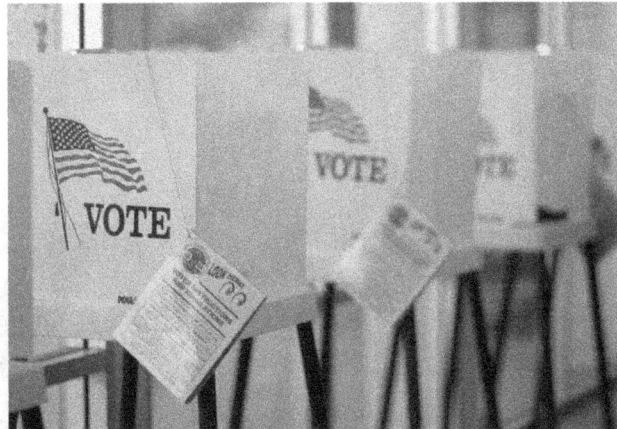

2. Suppose you take your answer to Question 1 to the candidate, who decides he wants a more precise poll. He offers to increase the budget so you can poll twice as many people. He thinks this will allow you to cut the size of the interval in half.

   Is he right? If you don't think so, explain why not, and explain how your poll size would need to change to cut the size of the interval in half.

A Matter of Confidence

Unless a pollster talks to every person in an overall population, there's no guarantee that poll results will give a true picture of the population. Therefore, a crucial question for every pollster is, "How confident can I be that my poll results are 'fairly close' to the truth?"

You will soon see how to turn this idea of "confidence" into a precise mathematical concept.

Tonisha Williams and Sierena Parker compare their results from experimenting with different values of p to determine what happens to σ in each case.

Different p, Different σ

Suppose you take a 20-person poll and get a sample proportion $\hat{p} = .60$. Of course, you would expect that the true proportion will be fairly close to this, but you can't be sure (unless your sample of 20 people is almost the entire population).

When you take a poll, there's a 95% chance the sample proportion will be within two standard deviations of p, the true proportion. And if $\hat{p}$ is within 2σ of p, then p is within 2σ of $\hat{p}$.

The interval of size 2σ around $\hat{p}$ is called a 95% **confidence interval.** Unfortunately, the value of σ itself depends on p, so you can't find this interval if you don't already know p. In this activity, you will explore this dilemma.

For this activity, remember that $n = 20$ and $\hat{p} = .60$.

1. Suppose p is .55.
    a. What is the value of σ?
    b. What are the endpoints of the 95% confidence interval around $\hat{p}$?
    c. Is p within the 95% confidence interval?

2. Now suppose p is .80.
    a. What is the value of σ?
    b. What are the endpoints of the 95% confidence interval around $\hat{p}$?
    c. Is p within the 95% confidence interval?

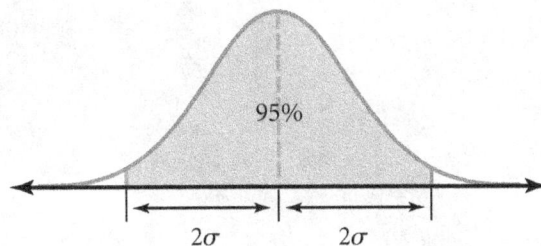

Let's Vote on It!

Is there an interesting issue about which you're curious how others feel? Here is your chance to ask some people—in fact, a whole "sample's worth" of people!

In this project, you and a partner will conduct a poll about a topic you choose. The topic must have only two sides, such as a yes-or-no issue, and should be something about which people are unlikely to be undecided.

Once you decide on a topic, you will select an overall population from which to choose a sample. Choose a population for which you can take a fairly random sample. For example, your population should not be "teenagers," because you can't possibly get a random sample of all teenagers. A more reasonable population would be "teenagers in our community."

You also need to decide how to choose your sample. You can't simply say, "Pick people randomly from the population." You have to give details. You also need to decide on a way to let respondents maintain their anonymity.

continued ▶

The last part of the planning involves deciding on the size of your poll. You will do this a little later in the unit, when you know more about the impact of sample size.

Eventually, you and your partner will turn in a write-up on your project. That write-up should include these items:

- A clear statement of the question voters are voting on
- How and why you selected your overall population
- The procedure you used to select people to poll in order to conduct a random sample of the population
- The procedure you used to ensure that people would vote honestly and the reason you think it was effective
- The size of your sample and the way you determined that number
- The results of your poll
- Your conclusions based on your results

In addition to a write-up, you and your partner will make a five-minute presentation to the class on your project. Plan to talk about the most interesting issues connected with your project.

Project Topics and Random Polls

Let's Vote on It! describes an important project for this unit. Although you will work on the polling project with a partner, in this activity you will work on your own. You may decide to change some of the decisions you make in this activity based on the discussion of the activity and on your partner's ideas.

In this activity, you will do these things:

• Choose a tentative polling topic and formulate a specific question

• Decide what your overall population will be

• Decide how you will choose a random sample from your population

• Decide how you will give participants enough anonymity so they will answer your question honestly

Mean, Median, and Mode

When you gather data, you often want to summarize the results using a single number to represent an "average" or "typical" result. Such a number is called a *measure of central tendency*.

The *mean*, which you have used throughout this unit, is one of the most commonly used measures of central tendency. Two others, you probably remember, are the *median* and the *mode*.

1. Review the precise meanings of *mean, median,* and *mode.* Look up the terms if necessary. Then state each of the definitions in your own words.

2. A student received grades on various assignments, with each grade based on a six-point scale. This graph shows how many times the student received each possible score.

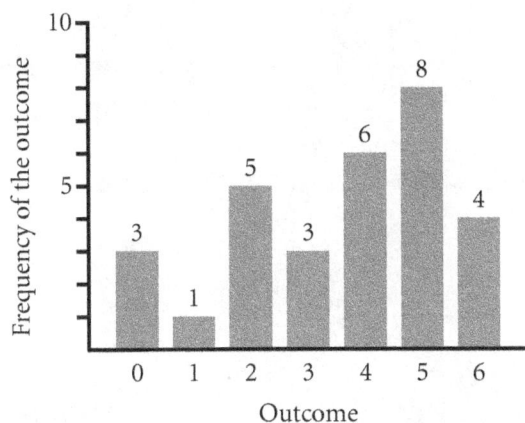

a. Find the mean, median, and mode for the data in the graph.

b. Discuss what each of these measurements tells you about the student's grades.

continued ▶

3. The measure of central tendency that is most meaningful varies from situation to situation.

   a. Make up a situation in which you think the *mean* would be the most meaningful measure of central tendency.

   b. Make up a situation in which you think the *median* would be the most meaningful measure of central tendency.

   c. Make up a situation in which you think the *mode* would be the most meaningful measure of central tendency.

The Worst-Case Scenario

You have seen that the reliability of a poll depends on σ, the standard deviation of the distribution of sample proportions. But $\sigma = \sqrt{\frac{p(1-p)}{n}}$, where p is the true proportion, so you can't find σ if you don't know p. And the whole purpose of a poll is to find out the value of p. So what do you do?

What professional pollsters generally do is consider the worst-case scenario. That is, they use the largest standard deviation that a poll of size n could have.

1. Pick a specific value for n. For that choice of n, experiment with different values of p and see what happens to σ in each case. Find the value of p that gives the largest value for σ.

2. Try other values of n, and generalize your results.

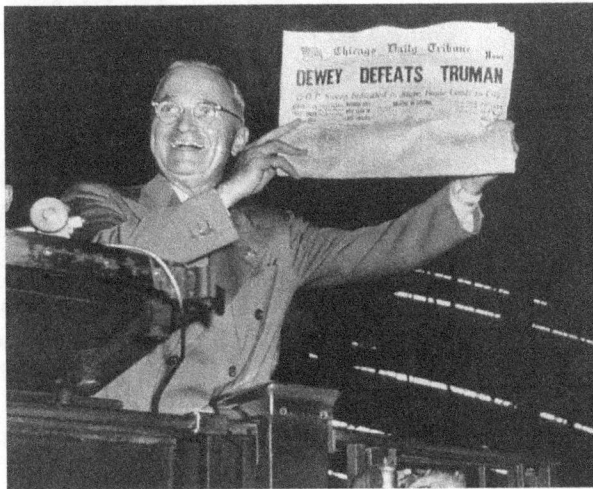

In 1948, President Harry S. Truman ran for reelection in a close race against New York governor Thomas E. Dewey. Although newspapers and pollsters widely predicted his defeat, Truman was delighted to find that the headline shown here was incorrect. He actually won the election by just over 2.1 million votes.

A Teaching Dilemma

Ms. Gordon is a new history teacher who is trying to develop some policies about grading. Her predecessor left her a copy of a test he had used for many years, along with the information that the mean test score had been 72 and the standard deviation had been 5.9.

Ms. Gordon has decided to use the test with her own students. She has no reason to think her students will perform differently from those in previous years. Assume her test scores are normally distributed, with a mean of 72 and a standard deviation of 5.9.

1. Suppose Ms. Gordon decides that students need to score above 85 to earn an A. What percentage of her students will get A's?

2. Suppose instead that Ms. Gordon decides she wants 3% of her students to earn A's. What should she set as the minimum score for an A?

3. Finally, suppose Ms. Gordon wants 80% of her class to earn C's. Suggest a reasonable range of scores for a grade of C that would probably contain about 80% of the test results. Is this the only possibility? Explain.

What Does It Mean?

River City is thinking of letting homeless people sleep in City Hall overnight.

Some people who live near City Hall are against the plan because they don't want homeless people gathering near their homes. People who favor the plan point out that there are homeless shelters in other neighborhoods. They say there is no reason why people living near City Hall should be allowed to avoid addressing the problem.

Other people oppose the plan because they think the city should do more than simply let people sleep in City Hall. They believe the city should provide housing with amenities, such as showers and cooking facilities. People who favor the plan counter by arguing that letting people sleep in City Hall is better than leaving them on the streets and that no better facilities seem to be available.

The local newspaper reports that it took a random telephone poll of registered voters. According to the poll, 52% favor the plan, with a margin of error of plus or minus 4%. Assume this margin of error is based on a 95% confidence level.

1. What is the confidence interval for this poll?

2. Explain what it means to say that the poll has a 95% confidence level.

3. Approximately how many people were in the sample?

4. Based on this result, how confident can one be that more than 50% of the voters favor this plan?

Confidence and Sample Size

1. The idea of a confidence interval is crucial for working with polls and other kinds of sampling.

    a. In your own words, explain what a confidence interval is.

    b. Give specific examples of how the concept of a confidence interval might be used and how it is related to the concepts of confidence level and margin of error.

2. Jen and Ken are having one of their famous debates.

    Jen claims that if you are sampling a bigger population, you need to take a bigger sample. In particular, she claims that taking a 500-person sample makes sense if your overall population is 40,000 people. But, she says, this is much too small a sample for a place like River City, with 400,000 voters. Ken doesn't think the size of the overall population matters at all.

    Would you please clear up their debate for them?

Polling Puzzles

Every pollster would like to have a high level of confidence and a small margin of error. And every pollster would love to be able to achieve this without polling a lot of people.

Unfortunately, these goals are incompatible. Confidence level, margin of error, and sample size are interconnected. As you will discover in this activity, you can control two of these three parameters at a time, but not all three.

1. Suppose a pollster wants a 2% margin of error with a 95% confidence level. How many people should he poll?

2. Suppose a pollster wants a 5% margin of error with a 97% confidence level. How many people should she poll?

3. Suppose a pollster wants to poll 400 people and have a 95% confidence level. What will the margin of error be?

4. Suppose a pollster wants to report a 3% margin of error with a 100-person poll. What will the confidence level be?

How Big?

You and your partner should have created a plan for how to choose people to poll in your *Let's Vote on It!* project. Now it's time to decide how many people to sample.

Discuss this issue with your partner. Think about how reliable and how precise you want your result to be. That is, what confidence level do you want, and what margin of error do you want? Keep in mind that if you are too demanding about these conditions, your sample size may become unmanageable.

Based on your decisions, find an appropriate sample size. Explain the calculations you did to come up with that number. This number and your explanation will be part of your final write-up for your project.

In addition, you will turn in a statement that includes these items:

- Your partner's name
- A statement of the topic for your poll
- A description of your overall population
- A discussion of how you will choose a random sample from your population and how you will give voters anonymity so that they will vote honestly
- The precise question you will ask participants

Putting It Together

You now have all the necessary tools to make a report to Coretta telling her what that original 53% poll result means. Before you do so, you will take a look back at a Year 2 problem about a suspicious coin.

You will then focus on your report and finalize your work for the *Let's Vote on It!* project.

Khareasha Stokes and Alexes Link listen to a presentation on the unit project.

Roberto and the Coin

In the Year 2 unit *Is There Really a Difference?*, you considered the problem of Roberto and his brother's coin. Here is the situation:

> Every time there is an extra dessert at Roberto's house, his older brother takes out his special coin. He always let Roberto flip the coin, and he always calls out "heads."

> Roberto believes that his brother wins more often than he should with a fair coin. One day, Roberto finds the coin and flips it 1000 times. He gets 573 heads and 427 tails.

Your task when you first encountered this problem was to decide whether the coin is fair. That meant figuring out how likely it would be for a fair coin to produce such unbalanced results. You used the χ^2 statistic to see that these results would be extremely unlikely with a fair coin—meaning Roberto had reason to be suspicious of his brother's coin. Now you will use different mathematical tools to explain what's going on.

1. Although the situation of coin flips involves the binomial distribution, the analysis can be approximated by the normal distribution.

   a. Explain what it means to say that the binomial distribution can be approximated by the normal distribution.

   b. Use the normal approximation to find out how likely it would be for a fair coin to give results as unbalanced as 573 heads and 427 tails.

2. Based on Roberto's experience with his brother's coin, give 95% and 99% confidence intervals for the true probability of getting heads with that coin.

3. Do you believe this is a fair coin? That is, do you think the apparent preference for heads is because the coin is not balanced, or is it simply a coincidence? Explain your reasoning.

How Much Better Is Bigger?

Is it always better to get a larger sample? In this activity, you'll explore how changing the poll size can affect either the confidence level or the margin of error.

1. Pick three large, but quite different, poll sizes. Assume you want a margin of error of 5%.

   a. For each poll size, find the confidence level you would have for that margin of error.

   b. Describe how the confidence level changes as the poll size increases.

2. Again, pick three different large poll sizes. (They can be the same as in Question 1.) This time, assume you want a 95% confidence level.

   a. For each poll size, find the margin of error you would have for this confidence level.

   b. Describe how the margin of error changes as the poll size increases.

3. Choose one of your examples, and explain what your values for the confidence level and the margin of error mean in that context.

The Pollster's Dilemma Revisited

It's time to return to the original unit problem, which involves Coretta's campaign for mayor of River City.

Recall that a poll was taken before you joined the campaign. That poll surveyed 500 people and found that 53% of them intend to vote for Coretta.

The question you need to answer is this:

What does this poll result mean?

Write a report for Coretta explaining how confident she should be that she was leading at the time of the poll. Explain your answer in terms of confidence intervals, standard deviation, and any other ideas you consider appropriate.

Final Data Collection

You will spend most of the next class period working on your project. You now need to complete your polling so that you will be prepared to analyze your results and finish your report.

The Pollster's Dilemma Portfolio

You will now put together your portfolio for *The Pollster's Dilemma.*
This process has three steps:

* Writing a cover letter summarizing the unit
* Choosing papers to include from your work in this unit
* Discussing your personal mathematical growth in the unit

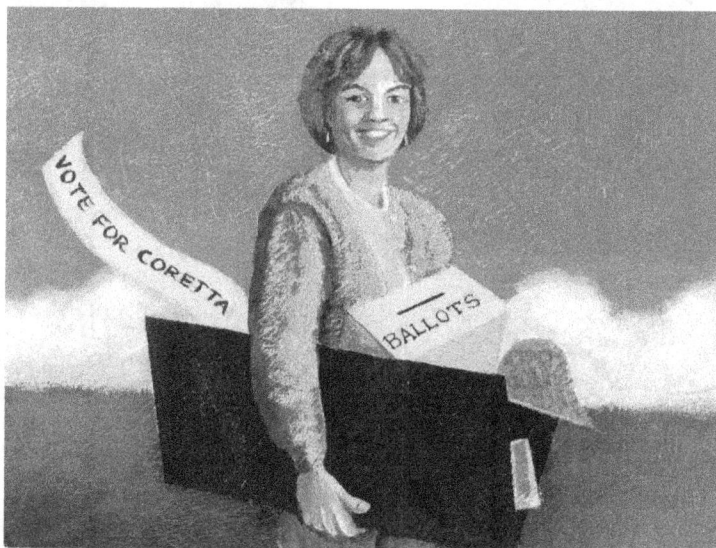

Cover Letter

Look back over *The Pollster's Dilemma,* and describe the central
problem of the unit and the main mathematical ideas. Your description
should give an overview of how the key ideas were developed in this
unit and how they were used to solve the central problem.

In compiling your portfolio, you will be selecting some activities that
you think were important in developing the key ideas of this unit. Your
cover letter should include an explanation of why you selected those
particular items.

continued ▶

Selecting Papers

Your portfolio from *The Pollster's Dilemma* should contain these items:

- An activity that shows the relationships among confidence interval, margin of error, and poll size
- An activity involving standard deviation
- POW 5: *The King's Switches*
- Your work on the unit project, *Let's Vote on It!*

Personal Growth

Over the past four years, you have had several units that dealt with statistics and probability. What have you learned? What general themes have come up repeatedly? How are concepts from statistics and probability used?

Summarize your experiences and your knowledge about statistics and probability. You don't have to include specific activities from units, but you should discuss your growth over time in your understanding of these two important branches of mathematics.

SUPPLEMENTAL ACTIVITIES

The supplemental activities for *The Pollster's Dilemma* continue the unit's focus on sampling and related ideas. Here are some examples:

- *What Is Random?* and *Random Number Generators* involve the important concept of randomness.

- *Three-Person Races* and *Another View of the Central Limit Theorem* concern generalizations of key ideas in the unit.

What Is Random?

One of the basic assumptions of the central problem in this unit is that the voters in the poll represent a random sample from the population of all voters.

The word *random* comes up in many contexts in mathematics. Your task in this activity is to summarize what you have learned about this concept.

You may find it helpful to review the situations in which you've used randomness. You might also consult a textbook or Web site on the theory of probability.

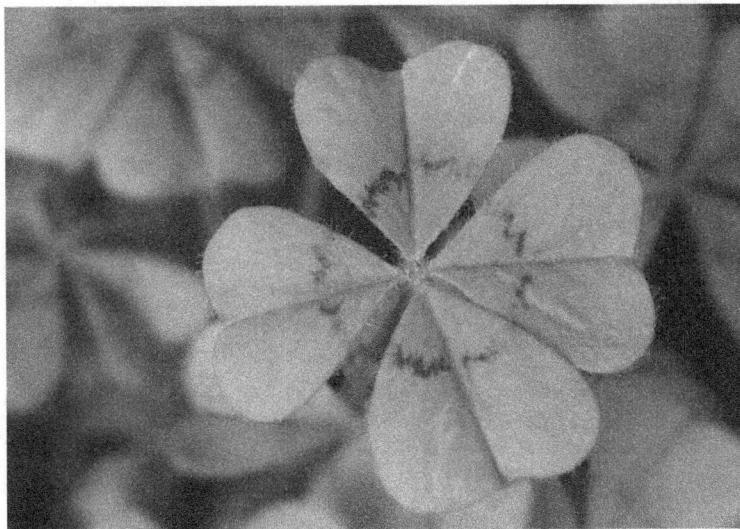

Random Number Generators

Many graphing calculators and computers have a random number generator. As you know, this is a device that supposedly picks numbers at random. Using a random number generator allows you to conduct various probability simulations.

The manufacturer of a graphing calculator or computer must find a way for the machine to generate random numbers, and the method may be different for different machines. Your task in this activity is to research and report on how a random number generator really works.

The Tack or the Coin?

You may remember a situation from the Year 2 unit *Is There Really a Difference?* that concerned Roberto, Roberto's brother, and a suspicious coin. Roberto's brother was using a coin to decide who should get the extra dessert when there was one.

Roberto isn't sure whether his brother's coin is fair, but he certainly doesn't like the results he has seen in the past. He tells his brother that the fate of the extra dessert will no longer be determined by that coin.

Roberto's brother chuckles and gives Roberto a new option, which works like this. They would take ten thumbtacks, shake them, and drop them on the ground.

• If more than five of the tacks land point up, Roberto will get the extra dessert.

• If fewer than five tacks land point up, Roberto's brother will get the extra dessert.

• If exactly five tacks land point up, the brothers will drop the tacks again.

1. Play around with Roberto's brother's new option and see what happens. Record your results in an organized way.

2. Choose another sample size for the bunch of tacks, and repeat the experiment over and over until you think you have a feel for how often tacks land point up. Again, record your results in an organized way.

3. Predict about how many tacks would land point up if you shook and then dropped 100 tacks on the ground.

4. Test your prediction for 100 tacks, and write about how it turned out.

5. Discuss how the number of tacks in the experiment and the number of times you do the experiment influence the reliability of your prediction.

Three-Person Races

In the probability analysis you did for the central unit problem, you assumed there were only two candidates. Of course, many elections have more than two candidates, and the mathematical analysis becomes more complicated. In this activity, you will examine the case of a three-person race.

Suppose there is an election race with three candidates: A, B, and C. Assume that in the overall population, the candidates have these percentages of support:

• Candidate A—40%

• Candidate B—35%

• Candidate C—25%

These numbers are called the *true proportions*. Notice that the true proportions add to 100%.

1. Suppose you select two people at random and ask them which candidate they are supporting.

   a. List the possible outcomes for the number of votes each candidate could get in such a poll.

   b. Find the probability of each possible outcome. Assume the overall population is large enough that you can treat your polls if it involves sampling with replacement. Check your work verifying that your probabilities add up to 1.

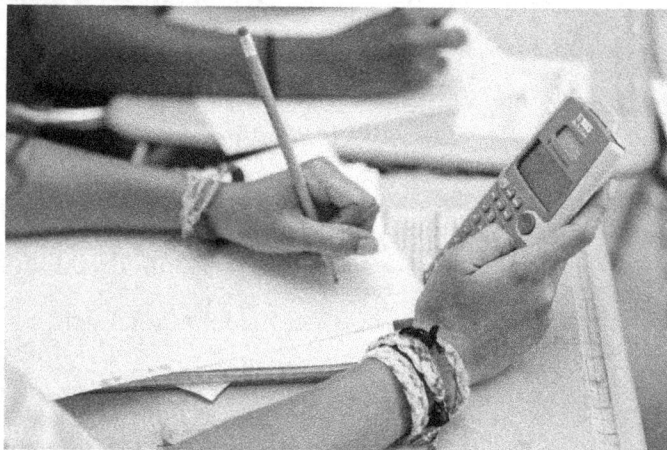

continued ▶

2. Repeat Question 1 for a poll of three people.

3. Generalize your results to a poll of n people. That is, find a formula for the probability that such a poll will have r votes for Candidate A, s votes for Candidate B, and t votes for Candidate C (where $r + s + t = n$).

You've seen that for polls involving only two candidates, the probabilities for various poll results can be expressed using combinatorial coefficients and powers of the true proportions. Build on your work in Questions 1 and 2 to see how to adjust for the case of three candidates, using something similar to combinatorial coefficients.

Generalizing Linear Interpolation

Gaps in the Table describes the technique known as *linear interpolation* for finding in-between values for functions. Your task in this activity is to develop a general formula for linear interpolation.

Assume that you have a function g and that you know the values of $g(a)$ and $g(b)$. Also assume that c is a number between a and b.

What would you use as an estimate for the value of $g(c)$, based on linear interpolation?

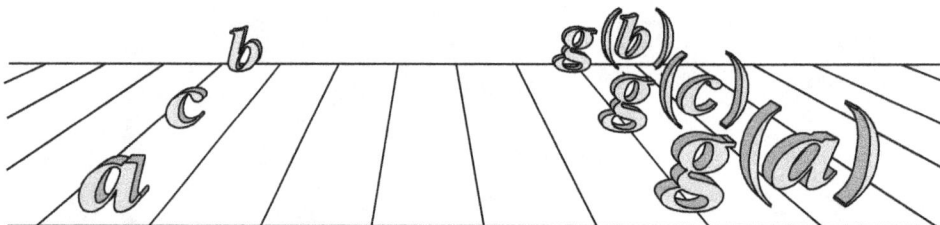

Another View of the Central Limit Theorem

This unit has focused on a special case of the central limit theorem. This special case says, in brief, that if n is "large enough," the normal distribution can be used to approximate the binomial distribution.

In this activity, you will examine a case of the central limit theorem in which the original distribution is not binomial.

Consider the spinner shown here, for which the probability of spinning 1 is .3, the probability of spinning 3 is .5, and the probability of spinning 9 is .2.

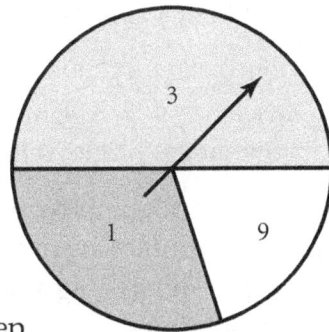

1. Make a probability bar graph for the outcomes of this spinner.

   The situation becomes more complex as you spin the spinner more and more times and keep track of the results. For instance, if you spin this spinner twice, you might get two 1s, a 1 and a 3, a 1 and a 9, two 3s, a 3 and a 9, or two 9s.

Now you'll consider different cases of n spins (your graph in Question 1 was the case $n = 1$) and see what happens as n increases.

2. Consider the case $n = 2$. That is, find the probability of each of the possible two-spin results.

To analyze the results of many spins, it's better to focus on the *average* result than on the *sum* of the results. (This is analogous to looking at the *proportion* of votes a candidate gets rather than the *number* of votes.) For instance, if you spin a 1 and a 9, your average result is 5.

3. a. Convert your results from Question 2 to averages. (If $n = 1$, the average is the same as the sum, so you don't need to convert the single-spin results.)

continued ▶

b. Make a probability bar graph of the results from Question 3a, using the same scale as in Question 1. You may want to redo your graph from Question 1 using a scale that works for both $n = 1$ and $n = 2$.

4. Consider the cases $n = 3$, $n = 4$, and $n = 5$. Find the probability distribution for the average result in each case, and make a bar graph of the results in each case, again using the same scale. (*Note:* Some averages can be achieved in different ways. For instance, for $n = 5$, you can get an average of 2.6 by spinning four 1s and one 9 or by spinning one 1 and four 3s. Either outcome results in a sum of 13, for an average of 2.6.)

5. Compare the probability bar graphs for the cases $n = 1, 2, 3, 4$, and 5. Describe how the graphs are changing.

6. Suppose you approximate the distribution of average results for 100 spins using a normal distribution.

a. What would you expect for the mean of the normal distribution? Explain your answer.

b. What would you expect for the standard deviation of the normal distribution? Explain your answer.

It's the News

The central problem of this unit concerns election polls. Polls are used in many situations to get information about what people think and do. Such polls appear in news stories regularly, although reports on the polls sometimes don't give as much information as they should.

Find a news article that reports on a poll.

When you've found an article, summarize what the report says. Discuss any shortcomings or weaknesses you see in the report. In particular, comment on any information that you think should have been included that would have helped you understand better what conclusions you could draw from the poll.

PHOTOGRAPHIC CREDITS

Front Cover Photography

(upper row) Stephen Loewinsohn; (lower left) Digital Images/ Getty Images; (background and lower right) Shutterstock; (lower middle right) iStockphoto

The Pollster's Dilemma

1 (upper row) Stephen Loewinsohn; (lower left) Digital Images/Getty Images; (background and lower right) Shutterstock; (lower middle right) iStockphoto; **3** Chicha Lynch, Hillary Turner, Richard Wheeler; **10** Shutterstock; **13** Stephen Loewinsohn; **20** Chicha Lynch, Hillary Turner, Richard Wheeler; **26** iStockphoto; **28** Stephen Loewinsohn; **33** iStockphoto; **39** Stephen Loewinsohn; **45** iStockphoto; **46** iStockphoto; **47** iStockphoto; **50** iStockphoto; **51** iStockphoto; **52** Stephen Loewinsohn; **58** iStockphoto; **59** Time & Life Pictures/Getty Images; **60** iStockphoto; **63** iStockphoto; **64** Stephen Loewinsohn; **65** Stephen Loewinsohn; **73** iStockphoto; **74** Shutterstock; **75** Shutterstock; **76** Stephen Loewinsohn